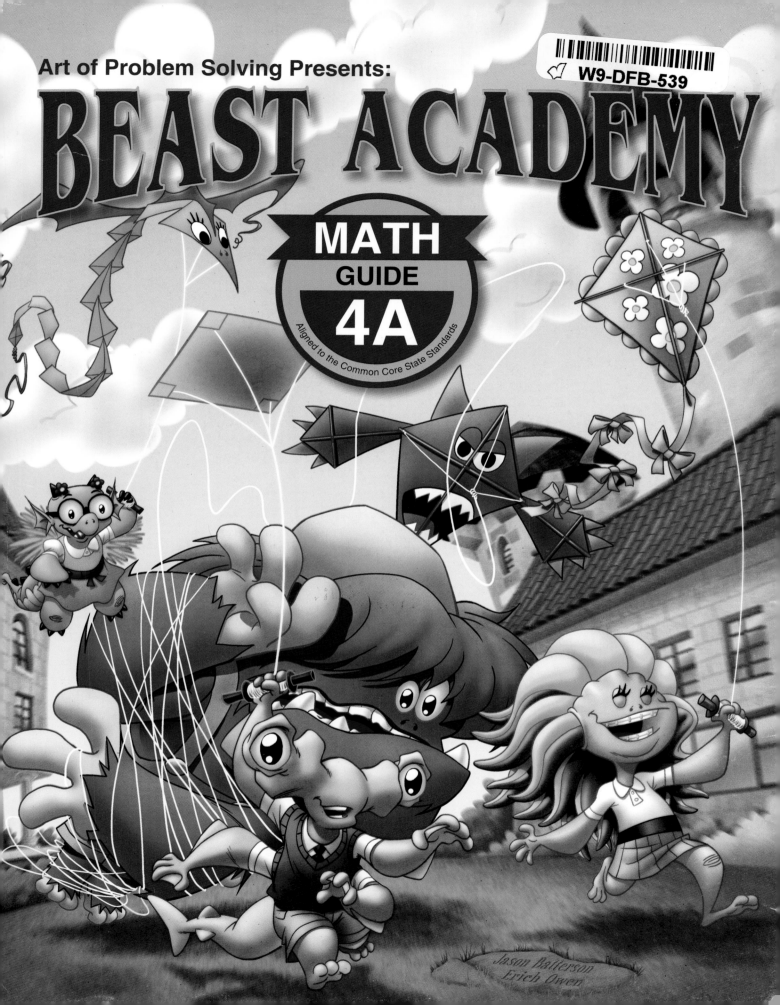

Published by: AoPS Incorporated
                   P.O. Box 2185
                   Alpine, CA 91903-2185
                   (619) 659-1612
                   info@BeastAcademy.com

ISBN: 978-1-934124-50-5

Beast Academy is a registered
trademark of AoPS Incorporated.

Written by Jason Batterson
Illustrated by Erich Owen
Colored by Greta Selman
Cover Design by Lisa T. Phan

Visit the Beast Academy website at www.BeastAcademy.com.
Visit the Art of Problem Solving website at www.artofproblemsolving.com.
Printed in the United States of America.
First Printing 2013.

Become a Math Beast!
For additional books,
printables, and more, visit
www.BeastAcademy.com

This is Guide 4A in a four-book series for fourth grade:

Contents:

The Headmaster
How to use this book

Welcome to Beast Academy!

This book is called the Guide.

There is also a separate Practice book with lots of problems you can use to sharpen your skills.

The Guide is written like a comic book.

In a comic book, whatever I say shows up in these bubbles. They're called comic balloons.

Here's one!

Each character has a different balloon color. This makes it easy to tell who is talking.

My balloons are purple!

The story is told in panels.

Panels usually have a rectangular frame around them...

...like this one.

But sometimes, panels don't have frames...

...like the open panel I'm in now.

If you've gotten this far, you probably know a little bit about how to read a comic book.

You read a comic book the same way you read any other book... from left to right and from top to bottom.

On each page, start in the top left panel.

Go to the right, then down.

Read all of the balloons in each panel from left to right and top to bottom before moving to the next panel.

At first, you may need to think about which balloon to read next.

Like when lots of characters are talking.

Or when a character speaks more than once.

Right! And sometimes several balloons get connected.

With a little practice, reading comics becomes natural.

How many panels are on this page?

Practice: Pages 6, 38, and 70

Contents: Chapter 1

See page 6 in the Practice book for a recommended reading/practice sequence for Chapter 1.

Chapter 1:
Shapes

R & G Definitions

What are you working on?

Ms. Q. left this in the copier. It's tonight's geometry homework.

You brought *homework* on our *fishing trip!?!*

I wanted to give it a try.

Let's see what you have so far.

Most of these words are pretty easy.

Define each of the following:

Point: What you do when you want to show someone something

Line: The string you tie to your fishing hook

Segment: part of a worm

Ray: A drop of golden sun

Angle: A fancy word for fishing

Plane: A flying vehicle

I don't think *these* are the definitions Ms. Q. is looking for.

Huh?

These are all geometry words.

Ms. Q. is probably looking for *math* definitions.

Let's look these up.

You brought your *Smartbot* on our *fishing trip!?!*

It has a high-resolution sonar-imaging fish finder!

What's the math definition of point?

"A *point* is an exact location."

Got it!

Wait, there's more!

"A point has no size, only position."

Huh?

I think it means that a point is a location, not a thing...

...like the center of a circle.

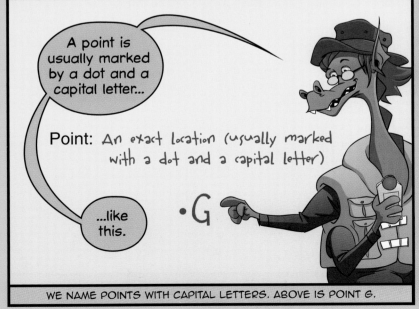

A point is usually marked by a dot and a capital letter...

Point: An exact location (usually marked with a dot and a capital letter)

...like this.

•G

WE NAME POINTS WITH CAPITAL LETTERS. ABOVE IS POINT G.

What's the next word?

Line.

"A *line* is a straight path that goes forever in both directions."

Forever?

That's what it says.

What's at the end?

There isn't an *end*. A line goes *forever.*

If a line goes forever, how do you draw it?

We usually put arrows on it.

So you know it keeps going.

Right.

A line is usually labeled with two points.

Line: A straight path that goes forever in both directions

A ⟷ B

A LINE IS NAMED BY TWO OF ITS POINTS. THIS IS LINE AB, OR LINE BA.

The next word is *segment*. What's a segment?

Part of a line from one point to another is called a *segment.*

Like this?

Exactly.

The points at the ends are called *endpoints.*

Segment: part of a line from one point to another

K ——— L

ABOVE IS SEGMENT KL, WHICH CAN ALSO BE NAMED SEGMENT LK.

The part of a line that starts at one point and goes forever in one direction is called a **ray**.

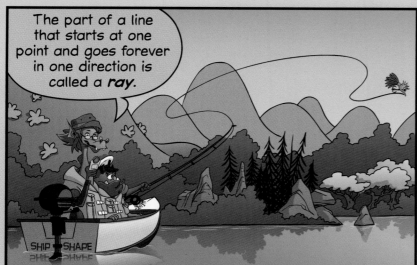

The point where the ray starts is called the **origin**.

Got it.

Ray: part of a line beginning at one point and extending forever in one direction

X   Y

WHEN NAMING A RAY, ALWAYS BEGIN WITH ITS ORIGIN. THE RAY ABOVE IS RAY XY, NOT RAY YX.

What's the math definition of **angle?**

"Two rays that start at the same point make an angle."

I remember now! A **right** angle makes a perfect "L".

That's right.

That's what I just said.

I mean, that's **correct!**

How's this?

Looks good.

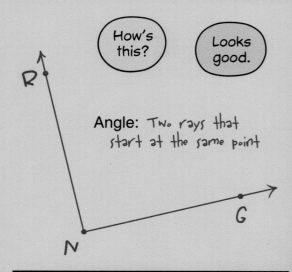

R

Angle: Two rays that start at the same point

G

N

ABOVE IS ANGLE RNG.
THE SHARED ORIGIN ALWAYS GOES IN THE MIDDLE WHEN WE NAME AN ANGLE. SO, WE CAN ALSO NAME THE ANGLE ABOVE ANGLE GNR, BUT NOT ANGLE RGN OR ANGLE GRN.

Let's define the last word so I can start fishing.

Okay.

Look up **plane.**

Good! Angles are measured in degrees.

With a thermometer!?!

No, with this.

How do you take an angle's temperature with *that!?!*

Let's start over.

Temperature degrees are different from *angle* degrees.

We use a *protractor* to measure the number of degrees in an angle.

BEAST ISLAND
PRO TRACTOR PULL
INVITATIONAL

Not a pro tractor, a *protractor!*

0 10 20 30 40 50 60 70 80 90 100 110 120 130 140 150 160 170 180

A *protractor* is usually a half-circle, with degrees marked from 0 to 180 around the outside.

To measure an angle, place the center of the half-circle on the angle's vertex.

Each ray should cross a degree mark along the outside of the protractor.

Turn your protractor so that one ray crosses the zero degree mark.

The degree mark where the other ray crosses gives you the angle measure.

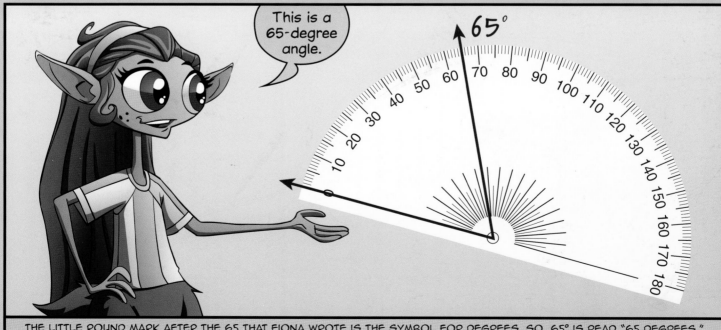

This is a 65-degree angle.

THE LITTLE ROUND MARK AFTER THE 65 THAT FIONA WROTE IS THE SYMBOL FOR DEGREES. SO, 65° IS READ "65 DEGREES."

I have an idea.

We can compare it to a right angle!

Every right angle has the same measure, 90 degrees.

I see.

Since this angle is acute, it must be less than 90°.

So, it can't be 152°, it must be 28°.

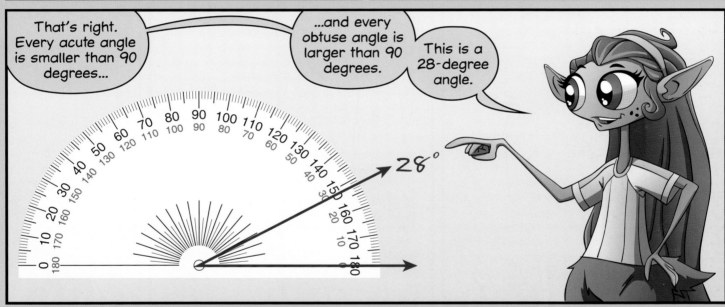

That's right. Every acute angle is smaller than 90 degrees...

...and every obtuse angle is larger than 90 degrees.

This is a 28-degree angle.

28°

Now that you can measure an angle, let's try drawing an angle with a given measure.

How would you draw a 103-degree angle?

We should start by drawing one ray.

The hard part is finding where the other ray should go.

How could you draw the other ray to make a 103-degree angle?

23

If we put the center of the half-circle on the vertex, and the first ray crosses through zero degrees...

...the second ray has to cross through 103 degrees.

We can draw a little dot at the 103.

Remember to use the 103 that makes the angle larger than 90 degrees.

Then, we draw a ray from the vertex through the dot!

There it is, a 103-degree angle.

This angle needs to see a doctor!

Huh?

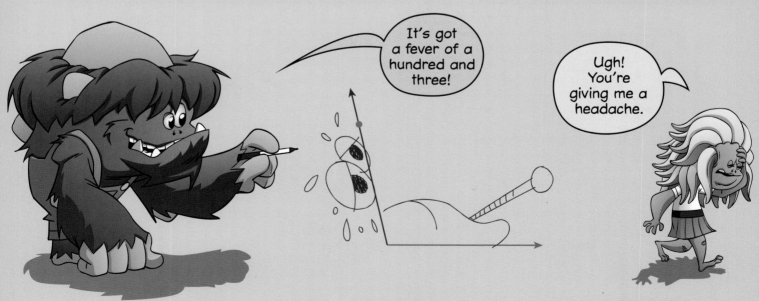

It's got a fever of a hundred and three!

Ugh! You're giving me a headache.

24

Practice: Pages 7-19

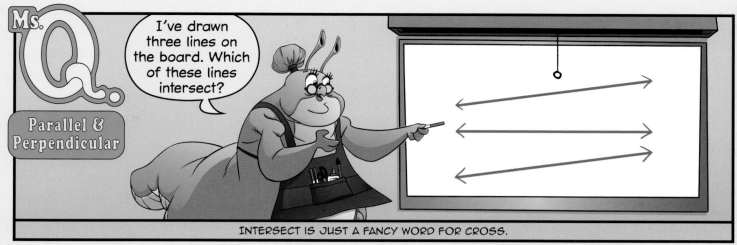

Ms. Q.
Parallel & Perpendicular

I've drawn three lines on the board. Which of these lines intersect?

INTERSECT IS JUST A FANCY WORD FOR CROSS.

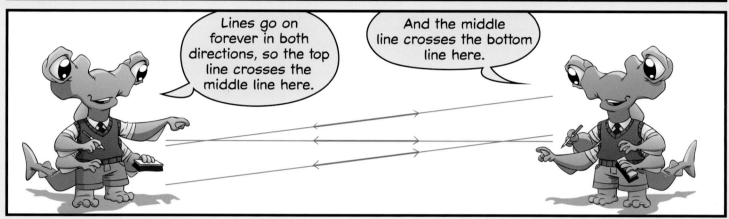

Lines go on forever in both directions, so the top line crosses the middle line here.

And the middle line crosses the bottom line here.

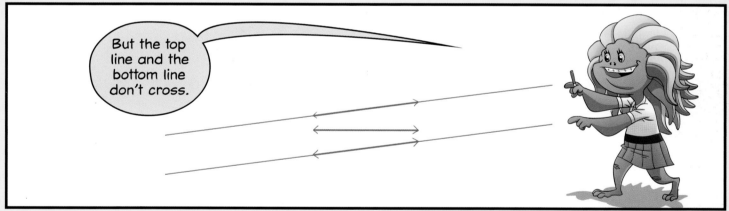

But the top line and the bottom line don't cross.

What do we call two lines that don't intersect?

"Two lines in the same plane that don't intersect are called *parallel* lines."

A PLANE IS AN ENDLESS FLAT SURFACE. FOR TWO LINES TO BE PARALLEL, THEY MUST BE IN THE SAME PLANE.

The sides of a rectangle that meet are perpendicular.

A right triangle has two sides that are perpendicular.

That's what makes it a right triangle.

This L-shaped hexagon has lots of pairs of perpendicular sides.

RIGHT ANGLES ARE OFTEN MARKED WITH SMALL SQUARES □ AS SHOWN IN THE DIAGRAMS ABOVE.

Good.

Can anyone guess what we call a quadrilateral in which both pairs of opposite sides are parallel?

Try to draw one.

squeak squeak squeak squeak

A Parallateral?

A Parallelogon?

A Parallelephant!

Grogg! That's not even a quadrilateral.

27

A quadrilateral in which **both** pairs of opposite sides are parallel is called a *parallelogram.*

Is every rectangle a parallelogram?

Opposite sides of a rectangle are parallel. If you extend them, they don't cross.

So, every rectangle is a parallelogram!

TWO SEGMENTS ARE PARALLEL IF THEY ARE PARTS OF LINES THAT ARE PARALLEL.

Perfect. Is every parallelogram a rectangle?

Nope.

This parallelogram isn't a rectangle.

And this rhombus is a parallelogram, but it's not a rectangle.

LEARN ALL ABOUT SQUARES, RECTANGLES, AND RHOMBUSES IN BEAST ACADEMY 3A.

Great!

Can a quadrilateral have just **one** pair of parallel sides?

Like this?

That's right! What do we call this shape?

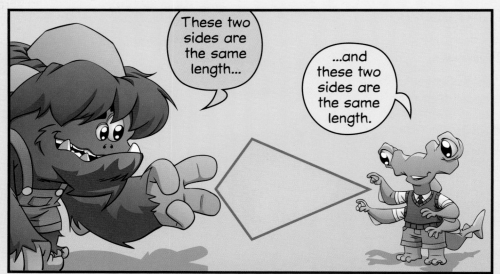

These two sides are the same length...

...and these two sides are the same length.

"A kite has two pairs of congruent sides that meet at opposite corners."

CONGRUENT SIDES ARE THE SAME LENGTH.

In a rectangle, the sides that are across from each other are the same length.

But in a kite, the sides that are the same length are next to each other.

That's right. Now, who wants make a kite?

Grogg!

Here be the most beautiful vessel ever sailed...

...the Helen of Ahoy.

What makes this ship so beautiful?

She be the picture o' perfect *symmetry*.

Huh?

The left side be exactly the same as the right.

If you slice the ship down the middle...

...one side o' the ship be a perfect *reflection* o' the other.

'Tis called *reflectional symmetry*.

A REFLECTION IS A MIRROR IMAGE.

If you can fold a shape so that one side matches the other side perfectly...

...then the shape has reflectional symmetry.

REFLECTIONAL SYMMETRY IS SOMETIMES CALLED MIRROR SYMMETRY OR LINE SYMMETRY.

Aye. The fold line be called a *line of symmetry.*

For example, this isosceles triangle be havin' exactly one line o' symmetry.

Some shapes be havin' more than one line o' symmetry.

How many lines o' symmetry be there in this equilateral triangle?

How many?

An equilateral triangle has *three* lines of symmetry.

Good work.

How about this rhombus? How many lines of symmetry be there in a rhombus?

How many?

Is this a line of symmetry?

How about this one?

Each of these lines splits the rhombus into two *congruent* pieces...

...but I don't think *those* are lines of symmetry.

Is Winnie right?

TWO FIGURES ARE *CONGRUENT* IF THEY ARE THE SAME SIZE AND SHAPE.

33

If you swap this purple patch with the green one down here...

...you get a sail with two lines of symmetry.

Aye. 'Tis a clever swap.

Shapes can also be havin' another type o' symmetry...

...rotational symmetry.

If you can turn a shape a different way without changing the way it looks, the shape has *rotational symmetry*.

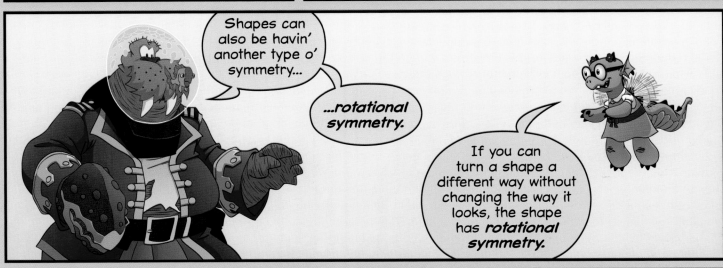

Aye. Take a close look at the compass rose on this map.

How many different ways can you turn the compass rose 'n' have it look the same?

THE LETTERS AROUND THE COMPASS ROSE ARE NOT CONSIDERED FOR ROTATIONAL SYMMETRY.

Try it.

When I turn it on its side, it looks exactly the same.

It looks the same upside-down, too!

Or turned this way!

We can turn it four different ways, and it still looks the same.*

Aye. The number o' ways you can turn a shape to look the same be called its **order** of rotational symmetry.

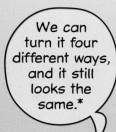

So, the compass rose has rotational symmetry of order 4.

Right you arrrr!

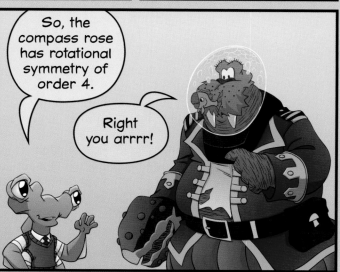

*THIS INCLUDES NOT TURNING IT AT ALL (WHICH IS THE SAME AS TURNING IT ALL THE WAY AROUND), PLUS THE THREE TURNS SHOWN IN THE PANELS ABOVE.

Who can find the symmetry in each o' these designs?

Which shapes have reflectional and/or rotational symmetry?

The dragon design has rotational symmetry of order 3.

The sword design has rotational symmetry of order 2.

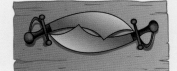

The skull doesn't have rotational symmetry, but it has one line of reflectional symmetry.

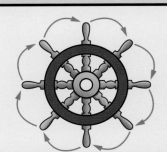

The ship's wheel has rotational symmetry of order 8...

...and 8 lines of reflectional symmetry!

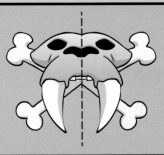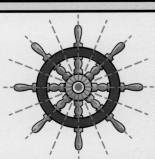

Well done!

Now I see why the ship is considered so magnificent.

Lucky for me, not everything has to be symmetric to be beautiful.

Practice: Pages 20-37

RECESS

Setup

The game is for two players.

Begin by folding a sheet of paper to divide it in half. Each player then traces five pirate ship outlines in ink on his or her side of the page. Players should use roughly the same five ship outlines. In our sample game, Grogg's ships are on the left and Winnie's are on the right.

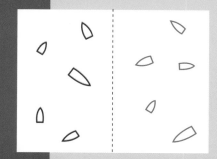

Attack!

Step 1:

Players take turns firing cannonballs. In the example, Grogg goes first by making a small, dark pencil mark on his side of the paper as shown. Here, Grogg is firing a cannonball at Winnie's smallest ship.

Step 2:

Grogg folds the paper in half. The dark pencil mark can be seen through the paper. Grogg scribbles over the mark, pressing on the back of the sheet to transfer his pencil mark to Winnie's side of the page.

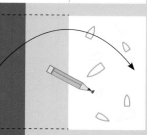

Step 3:

Grogg unfolds the paper to see where the cannonball landed. In the example shown, Grogg's shot was a miss. If the transferred mark touches your opponent's ship, the ship is sunk and marked with an X.

Victory!

The first player to sink all five opponent ships is the winner. Grogg is the winner of the game shown on the right.

Sample game boards, ship outlines, and game variations can be found at BeastAcademy.com.

Ambigrams by Grogg

Contents: Chapter 2

See page 38 in the Practice book for a recommended reading/practice sequence for Chapter 2.

Chapter 2:
Multiplication

It's like finding the area of a big rectangle by splitting it into smaller pieces.

We find the area of each piece, then add the pieces together to find the total area.

123

7

100   20 3

7

$7 \times 123 = 7 \times 100 + 7 \times 20 + 7 \times 3$
$= 700 + 140 + 21$
$= 861$

That's right!

Today we are going to learn how to use the distributive property to multiply two 2-digit numbers.

How could you use the distributive property to multiply 27×38?

One part at a time!

38

27

To multiply 27×38, we can split 38 into two parts: $30+8$.

27×38
$= 27 \times (30 + 8)$

Then, we multiply each part by 27.

27×38
$= 27 \times (30 + 8)$
$= 27 \times 30 + 27 \times 8$

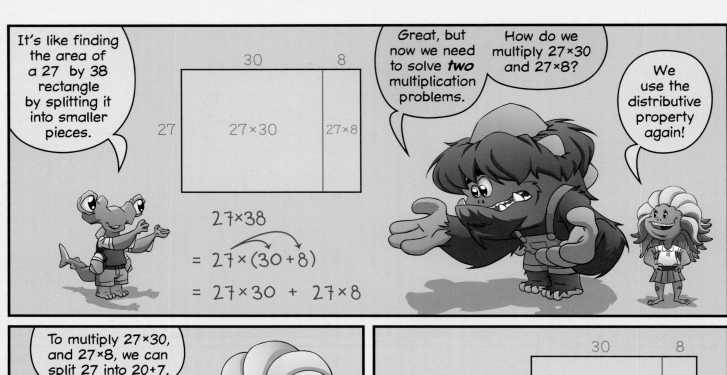

It's like finding the area of a 27 by 38 rectangle by splitting it into smaller pieces.

$$27 \times 38$$
$$= 27 \times (30 + 8)$$
$$= 27 \times 30 + 27 \times 8$$

Great, but now we need to solve *two* multiplication problems.

How do we multiply 27×30 and 27×8?

We use the distributive property again!

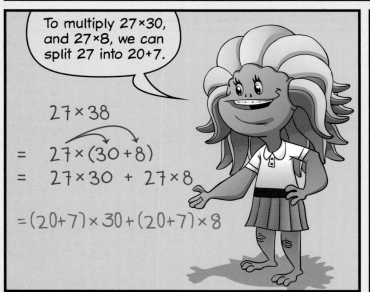

To multiply 27×30, and 27×8, we can split 27 into 20+7.

$$27 \times 38$$
$$= 27 \times (30 + 8)$$
$$= 27 \times 30 + 27 \times 8$$
$$= (20 + 7) \times 30 + (20 + 7) \times 8$$

We can do that on our area diagram, too!

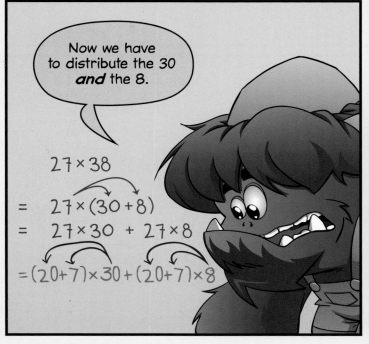

Now we have to distribute the 30 *and* the 8.

$$27 \times 38$$
$$= 27 \times (30 + 8)$$
$$= 27 \times 30 + 27 \times 8$$
$$= (20 + 7) \times 30 + (20 + 7) \times 8$$

It's the same as finding the areas of all four parts of our rectangle!

We get these four products.

And the same four products here!

	30	8
20	20×30	20×8
7	7×30	7×8

27×38

$= (20+7) \times (30+8)$

$= 20 \times 30 + 20 \times 8 + 7 \times 30 + 7 \times 8$

And to finish, we compute the products and add them!

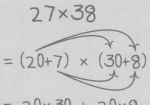

```
  600
  210
  160
+  56
------
 1026
```

	30	8
20	20×30 =600	20×8 =160
7	7×30 =210	7×8 =56

27×38

$= (20+7) \times (30+8)$

$= 20 \times 30 + 20 \times 8 + 7 \times 30 + 7 \times 8$

$= 600 + 160 + 210 + 56$

$= 1.026$

Marvelous!

We are using the distributive property to multiply two-digit numbers.

Who would like to try 53×46 without drawing a rectangle?

Today, I'm going to teach you how to multiply using an **algorithm.**

I've got rhythm!

Not rhythm, **algorithm.**

"An algorithm is a set of steps used for solving a problem."

Today's algorithm is for multiplying any two numbers.

Let's start with an example.

Multiply 6×458.

$$6 \times 458$$

We split 458 into three parts, multiply, then add the parts back together.

$$6 \times 458$$
$$= 6 \times (400 + 50 + 8)$$
$$= 6 \times 400 + 6 \times 50 + 6 \times 8$$
$$= 2400 + 300 + 48$$

$$\begin{array}{r} 2400 \\ 300 \\ + \quad 48 \\ \hline \underline{2748} \end{array}$$

$6 \times 458 = 2{,}748.$

Perfect!

Multiplying requires a **lot** of writing.

49

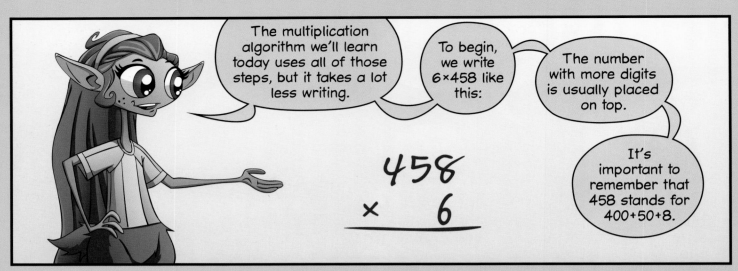

The multiplication algorithm we'll learn today uses all of those steps, but it takes a lot less writing.

To begin, we write 6×458 like this:

The number with more digits is usually placed on top.

It's important to remember that 458 stands for 400+50+8.

$$458 \times 6$$

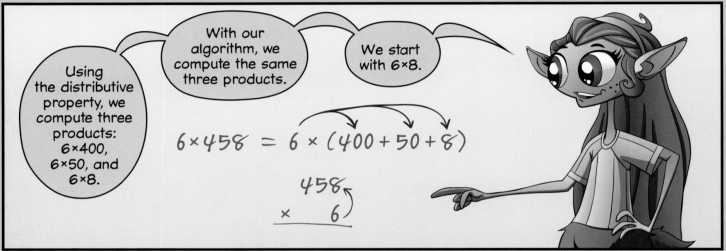

Using the distributive property, we compute three products: 6×400, 6×50, and 6×8.

With our algorithm, we compute the same three products.

We start with 6×8.

$$6 \times 458 = 6 \times (400 + 50 + 8)$$

$$458 \times 6$$

6×8 =48.

Good. We write 48 down here.

$$458 \times 6 = 48$$

The 5 in 458 stands for 5 tens, or 50.

We multiply 6×50 next.

$$458 \times 6 = 48$$

6×50 =300.

Right. We write 300 here. Line up the digits with the 48 above it.

$$458 \times 6 = 48 \; 300$$

DIGITS FOR EACH PLACE VALUE SHOULD LINE UP.

6 × 458

= 6 × (400+50+8)

= 6×400 + 6×50 + 6×8

= 2400 + 300 + 48

     2400
      300
 +   48
  (2748)

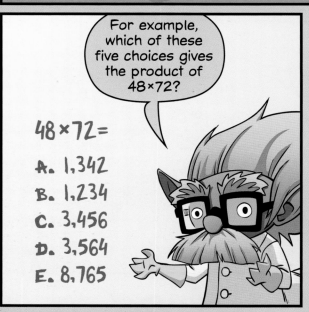

When we multiply 48×72, we split it into four products.

I see.

Three of the products are multiples of 10.

So these three products all have a zero in the ones place.

$$
\begin{array}{r}
48 \\
\times\ 72 \\
\hline
16 \\
80\ \ \leftarrow 2\times40 \\
560\ \ \leftarrow 70\times8 \\
+\ 2800\ \ \leftarrow 70\times40
\end{array}
$$

But for *this* product, we are multiplying the ones digits of both numbers. 2×8=16.

Then, when we add the ones digits, we get 6+0+0+0=6.

Right. To find the ones digit of 48×72, we only need to multiply 8×2.

$$
\begin{array}{r}
48 \\
\times\ 72 \\
\hline
16\ \ \leftarrow 2\times8 \\
80 \\
560 \\
+\ 2800 \\
\hline
3456
\end{array}
$$

The ones digit of 48×72 is the same as the ones digit of 8×2!

Very good! To find the *units digit* of a product, you can multiply the units digits of the numbers you are multiplying.

Try this one.

What is the units digit of 67×89?

$$
\begin{array}{r}
67 \\
\times\ 89 \\
\hline
\end{array}
$$

Try it!

THE NUMBER IN THE ONES PLACE OF A NUMBER IS ITS ONES DIGIT, AND IS ALSO CALLED ITS *UNITS DIGIT*.

It would take us forever to multiply just *five* 99's...

...how are we ever going to multiply *ninety-nine* 99's?

For 67×89, we only needed to multiply 7×9 to find the units digit.

We don't actually need to multiply ninety-nine 99's to find the units digit...

...we only need to multiply the units digits.

$99 \times 99 \times 99 \times 99 \times 99 \times \cdots$

Maybe if we find the units digit for a product of two 99's, then three 99's, then four 99's...

...we can keep going until we find a pattern.

Since 9×9=81, the units digit of 99×99 is 1.

When we multiply *two* 99's, the units digit is 1.

What if we multiply *three* 99's?

$\underbrace{99 \times 99}_{\text{ends in 1}} \times 99 \times 99 \times 99 \times \cdots$

Can you find a pattern?

LOOKING FOR A PATTERN IS A GREAT WAY TO START A DIFFICULT-LOOKING PROBLEM.

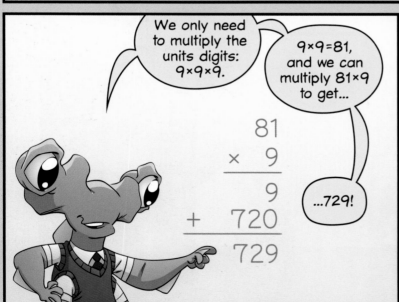

We only need to multiply the units digits: 9×9×9.

9×9=81, and we can multiply 81×9 to get...

$$\begin{array}{r} 81 \\ \times\ 9 \\ \hline 9 \\ +\ 720 \\ \hline 729 \end{array}$$

...729!

When we multiply three 99's, the units digit is 9!

Great, but you didn't need to multiply 81×9 to find the units digit!

Oh, right! The units digit of 81×9 is just 1×9=9!

So, 99×99 ends in 1, and 99×99×99 ends in 9.

	units digit
99×99	1
99×99×99	9
99×99×99×99	

What if we multiply *four* 99's?

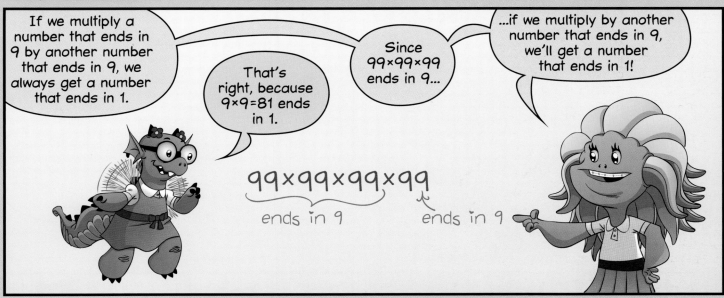

If we multiply a number that ends in 9 by another number that ends in 9, we always get a number that ends in 1.

That's right, because 9×9=81 ends in 1.

Since 99×99×99 ends in 9...

...if we multiply by another number that ends in 9, we'll get a number that ends in 1!

99×99×99×99

ends in 9   ends in 9

When we multiply four 99's, the units digit is 1.

	units digit
99×99	1
99×99×99	9
99×99×99×99	1
99×99×99×99×99	

Next is five 99's.

Four 99's give us a product that ends in 1. If we multiply by another 99, we get a number that ends in 9!

Because 1×9=9!

What is the units digit of the product of ninety-nine 99's?

59

Watch this!

You're not going to try to yank the tablecloth again, are you?

Huh?

Remember what happened last time?

No, it's nothing like that! I'm going to solve a multiplication problem!

I can multiply a 3-digit number by a 2-digit number!

Alright, let's see it.

Watch me multiply 456×91!

456
× 91

It's all about staying organized.

You start by multiplying the 1 in 91 by each digit in 456.

The 5 stands for 50, and the 4 stands for 400.

$$456 \times 91$$

6
50
400

1×6=6.
1×50=50.
1×400=400.

That gives you these three **partial products.**

456
× 91

6
50
400

THE PRODUCTS YOU GET IN EACH STEP ARE CALLED **PARTIAL PRODUCTS.**

Let's try another.

What do you get when you multiply 876×14?

Hmmm... First, I multiply the 4.

876
× 14
24
280
3200

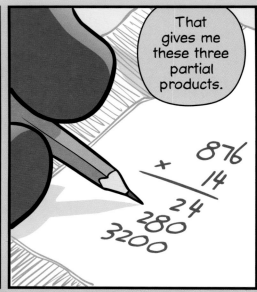

That gives me these three partial products.

876
× 14
24
280
3200

Then, I multiply the 1, which stands for 10.

I multiply 10×6, 10×70, and 10×800.

Right, but you--

Wait, I know!

876
× 14
24
280
3200

I can multiply 10×876 all at once!

10×876 =8,760.

876
× 14
24
280
3200
8760

So, 876×14 =12,264.

That's right!

When you multiply by 1, or by 10, or by 100, you don't need to multiply one part at a time.

876
× 14
24
280
3200
+ 8760
12264

Try this. What is 203×104?

Try it.

64

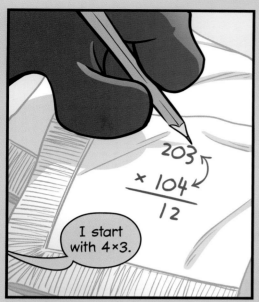

I start with 4×3.

Next comes 4×0.

The 0 stands for zero tens... which is still zero.

That's easy. 4×0=0.

I don't even need to write anything down...

...since adding zero doesn't change a sum.

Next comes 4×200. That's 800.

Now we move on to this 0.

That's easy again! Zero times anything is zero!

I can just move on to the 1.

It's in the hundreds place, so it stands for 100.

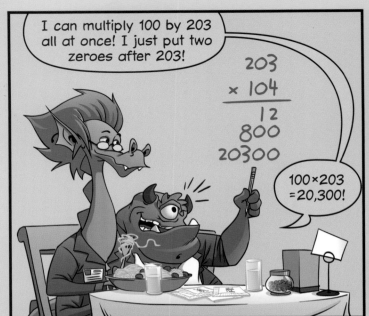

Contents: Chapter 3

See page 70 in the Practice book for a recommended reading/practice sequence for Chapter 3.

Chapter 3: Exponents

Congratulations!
You have reached the peak of

Mount Wanowanwanoh

Elevation:
10,110,010,111,011 feet
(110,110,100,010 meters)

Highest point on Binary Island

R & G
Exponents

$3^1 = 3$
$3^2 = 9$
$3^3 = 27$
$3^4 = 81$
$3^5 = $

Look at these tiny numbers.

Why are they so small?

Those little numbers are called **exponents**.

$3^1 = 3$
$3^2 = 9$
$3^3 = 27$
81

What are they for?

Exponents give us a shortcut for writing repeated multiplication.

Huh?

This 3 is the number you are multiplying. 3 is called the **base**.

3^4

The 4 is the exponent. It tells you how many 3's to multiply.

3^4

3^4 IS READ "3 TO THE 4TH POWER," OR SIMPLY "3 TO THE 4TH."

So, 3^4 means $3 \times 3 \times 3 \times 3$?

Exactly.

$$3^4 = 3 \times 3 \times 3 \times 3$$

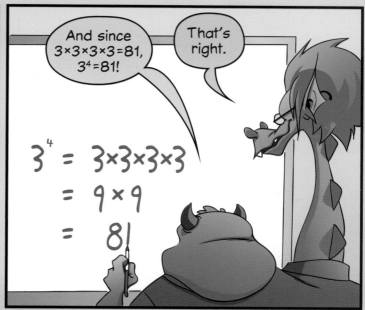

And since $3 \times 3 \times 3 \times 3 = 81$, $3^4 = 81$!

That's right.

$$3^4 = 3 \times 3 \times 3 \times 3$$
$$= 9 \times 9$$
$$= 81$$

Ms. Q. left 3^5 blank. Let's see if we can figure out what 3^5 equals.

$$3^1 = 3$$
$$3^2 = 9$$
$$3^3 = 27$$
$$3^4 = 81$$
$$3^5 =$$

Try it.

3^5 IS READ "3 TO THE 5TH POWER."

We know that $3^5 = 3 \times 3 \times 3 \times 3 \times 3$.

So, we need to multiply five 3's.

$$3^5 = 3 \times 3 \times 3 \times 3 \times 3$$

$$3^5 = 3 \times 3 \times 3 \times 3 \times 3$$
$$= 3^4 \times 3$$

Since multiplying **four** 3's gives us $3^4 = 81$.

To multiply **five** 3's, we can multiply 3^4 by one more 3!

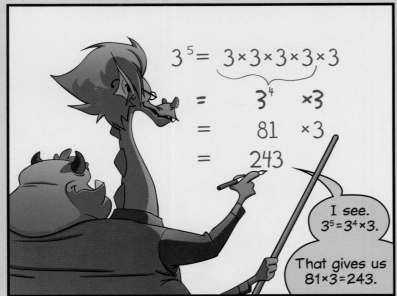

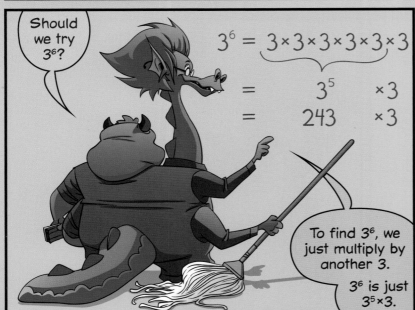

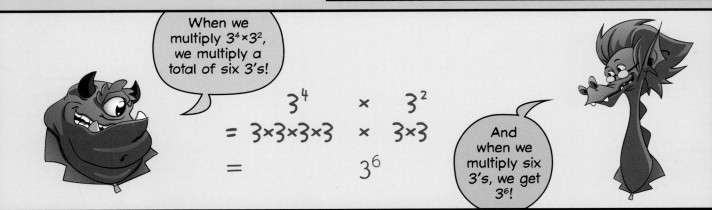

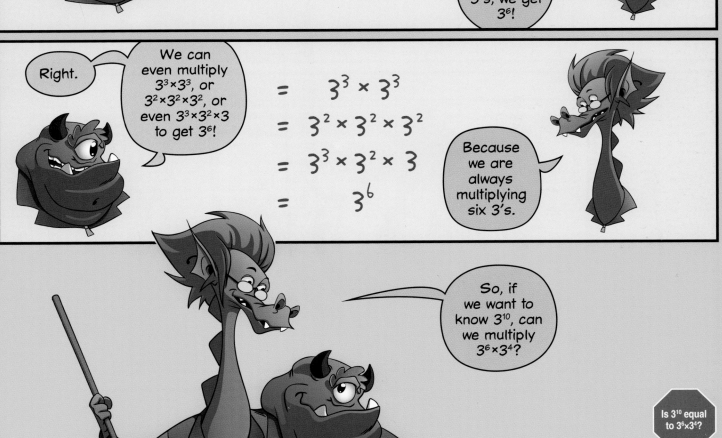

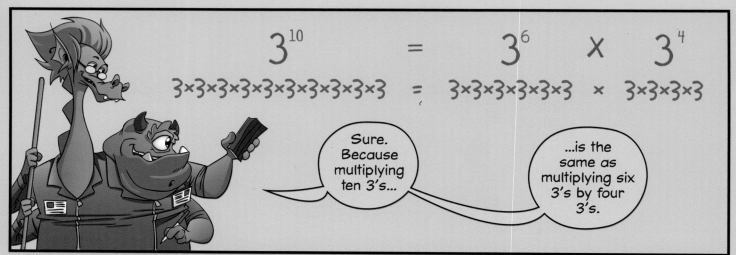

Lizzie

Powers

A power is a shortcut for writing repeated multiplication.
For example, we can write 5×5×5 as a power: 5^3.

$$5 \times 5 \times 5 = 5^3 \leftarrow \text{Exponent}$$
$$\hspace{4.5cm} \leftarrow \text{Base}$$

Every power has a base and an exponent.
The exponent tells us how many of the base we multiply.

5^3 is a power of 5 and is read, "five to the third power."

More examples:

$7 \times 7 \times 7 \times 7 = 7^4$ "seven to the fourth power"

$3 \times 3 = 3^2$ "three to the second power," or "three squared"

$2 \times 2 \times 2 \times 2 \times 2 \times 2 = 2^6$ "two to the sixth power"

Powers of 2:

$$2 = 2^1$$
$$2 \times 2 = 2^2 = 2^1 \times 2 = 4$$
$$2 \times 2 \times 2 = 2^3 = 2^2 \times 2 = 8$$
$$2 \times 2 \times 2 \times 2 = 2^4 = 2^3 \times 2 = 16$$
$$2 \times 2 \times 2 \times 2 \times 2 = 2^5 = 2^4 \times 2 = 32$$
$$2 \times 2 \times 2 \times 2 \times 2 \times 2 = 2^6 = 2^5 \times 2 = 64$$
$$2 \times 2 \times 2 \times 2 \times 2 \times 2 \times 2 = 2^7 = 2^6 \times 2 = 128$$
$$2 \times 2 \times 2 \times 2 \times 2 \times 2 \times 2 \times 2 = 2^8 = 2^7 \times 2 = 256$$
$$2 \times 2 \times 2 \times 2 \times 2 \times 2 \times 2 \times 2 \times 2 = 2^9 = 2^8 \times 2 = 512$$
$$2 \times 2 \times 2 \times 2 \times 2 \times 2 \times 2 \times 2 \times 2 \times 2 = 2^{10} = 2^9 \times 2 = 1,024$$

1. Evalute what's in parentheses

2. Multiply and divide from left to right

3. Add and subtract from left to right

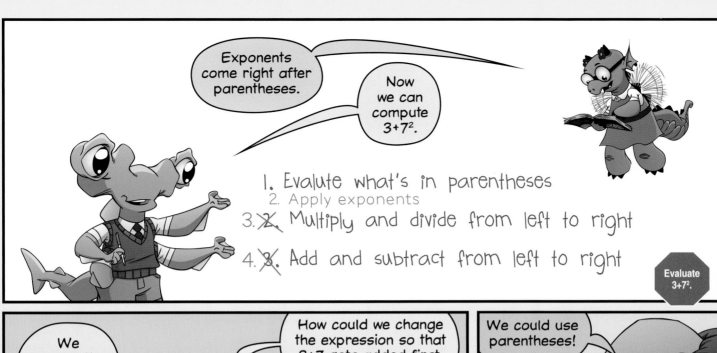

Exponents come right after parentheses.

Now we can compute $3+7^2$.

1. Evalute what's in parentheses
2. Apply exponents
3. ~~2.~~ Multiply and divide from left to right
4. ~~3.~~ Add and subtract from left to right

Evaluate $3+7^2$.

We square the 7 **before** we add.

$3+7^2$ =52.

How could we change the expression so that 3+7 gets added first, then squared?

Good!

$$3 + 7^2$$
$$= 3 + 49$$
$$= 52$$

We could use parentheses!

$$(3 + 7)^2$$
$$= 10^2$$
$$= 100$$

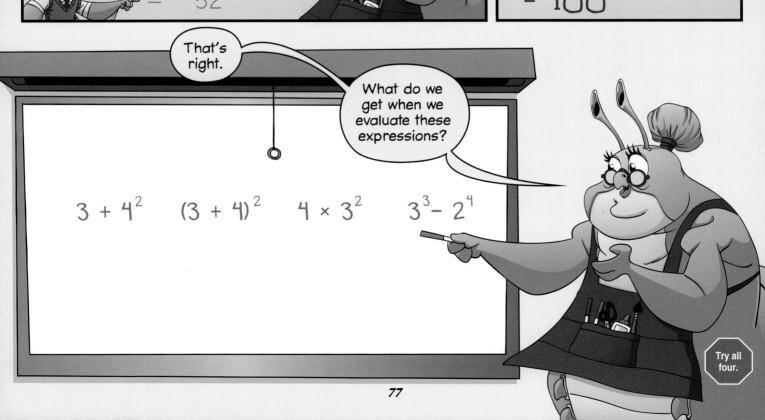

That's right.

What do we get when we evaluate these expressions?

$$3 + 4^2 \qquad (3 + 4)^2 \qquad 4 \times 3^2 \qquad 3^3 - 2^4$$

Try all four.

RECESS

Power Play is a game for two or more players. You will need a deck of cards numbered 0 through 9 (you can make your own, or print a set from www.BeastAcademy.com).

Each round, two cards are turned over from the top of the deck to create a two-digit target number.

Every player then attempts to write the target number as a sum of two or more powers.

For example, if the cards turned are a 5 then a 2, players attempt to write a sum of powers that is equal to 52. There are a few rules:

1. You may not use powers of 1.

$6^2 + 5^2 + 1^3 = 52$ is not allowed.

2. All exponents must be greater than 1.

$7^2 + 3^1 = 52$ is not allowed.

3. You may not use the same base more than once.

$2^5 + 2^4 + 2^2 = 52$ is not allowed.

There are three ways to write 52 as a sum of two or more powers using the rules above. They are listed upside-down at the bottom of this page.

Sometimes, the target number is impossible to write as a sum of powers. In this case, the goal is to write a sum of powers that is as close as possible to the target.

Scoring

Each game lasts 5 rounds. The goal is to have the **lowest** score after the final round.

Score each turn as follows:

- If your expression equals the target number, you get 0 points.
- If your expression does not equal the target number, your score is the difference between the value of your expression and the target number.

Add your score for each of the five rounds to get your total score. The winner is the player who scores the fewest points.

Variations

You may set a time limit for each round.

To make the game easier, you may eliminate rule 3 above.

Find a partner and play!

$6^2 + 2^4 = 52, \quad 6^2 + 4^2 = 52, \quad 5^2 + 3^3 = 52$

Let's explore what happens when you multiply two perfect squares.

Try $5^2 \times 3^2$.

$5^2 = 25$, and $3^2 = 9$, so $5^2 \times 3^2 = 25 \times 9$.

$$\begin{array}{r} 25 \\ \times \quad 9 \\ \hline 45 \\ + \quad 180 \\ \hline 225 \end{array}$$

$25 \times 9 = 225$!

That's weird. $5^2 \times 3^2 = 225$.

$0^2 = 0$	$7^2 = 49$	$14^2 = 196$
$1^2 = 1$	$8^2 = 64$	$15^2 = 225$
$2^2 = 4$	$9^2 = 81$	$16^2 = 256$
$3^2 = 9$	$10^2 = 100$	$17^2 = 289$
$4^2 = 16$	$11^2 = 121$	$18^2 = 324$
$5^2 = 25$	$12^2 = 144$	$19^2 = 361$
$6^2 = 36$	$13^2 = 169$	$20^2 = 400$

And 15^2 is also 225.

Cool. $5 \times 3 = 15$, and $5^2 \times 3^2 = 15^2$.

That's right!

Do we always get a perfect square when we multiply two perfect squares?

Let's explore!

Try $7^2 \times 2^2$.

$7^2 \times 2^2$

Is $7^2 \times 2^2$ equal to 14^2?

Try it.

81

$7^2 \times 2^2 = 49 \times 4$

$7^2 = 49$, and $2^2 = 4$, so $7^2 \times 2^2 = 49 \times 4$.

$49 \times 4 = 196!$

$$\begin{array}{r} 49 \\ \times\ \ 4 \\ \hline 36 \\ +\ 160 \\ \hline 196 \end{array}$$

And $14^2 = 196!$

It worked! $7^2 \times 2^2 = 14^2!$

$14^2 = 196$

That's right. Multiplying $7^2 \times 2^2$ gives the same result as multiplying 7×2, then squaring.

Who can explain why $7^2 \times 2^2 = (7 \times 2)^2$?

I have an idea.

$7^2 \times 2^2$
$= (7 \times 7) \times (2 \times 2)$

To compute $7^2 \times 2^2$, we multiply $(7 \times 7) \times (2 \times 2)$.

We can remove the parentheses, and rearrange the numbers.

$7^2 \times 2^2$
$= 7 \times 7 \times 2 \times 2$
$= 7 \times 2 \times 7 \times 2$

$7^2 \times 2^2$
$= 7 \times 7 \times 2 \times 2$
$= 7 \times 2 \times 7 \times 2$
$= (7 \times 2) \times (7 \times 2)$

Then, we can place parentheses here and here.

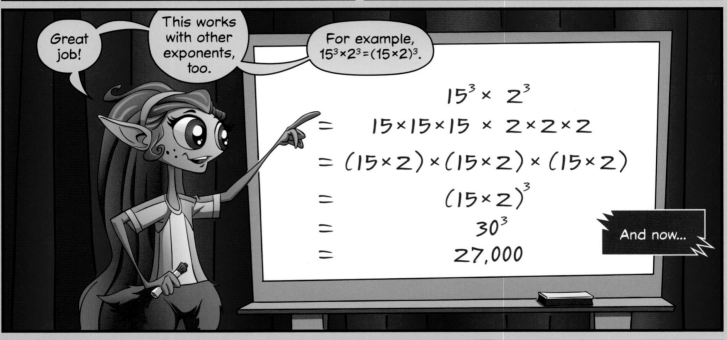

84

Exponents are the focus of today's math meet. I will ask 6 questions. The first five are each worth one point, and the final question is worth two. The team with the most points wins the meet. Is everyone ready for the first question?

Question 1: Compute 2^8.

Try it.

Ding! BZZZZT!

256.

256.

We'll have to go to the judges on that one.

The judges award both teams a point!

How did you get that so fast, Lizzie?

I learned all of the powers of 2 up to $2^{10} = 1,024$.

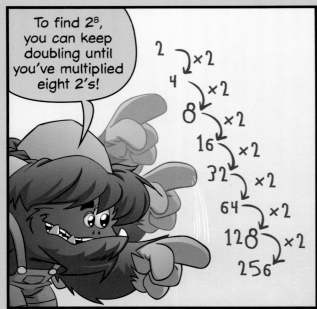

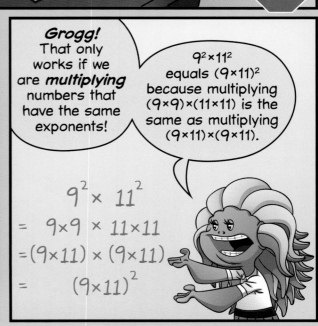

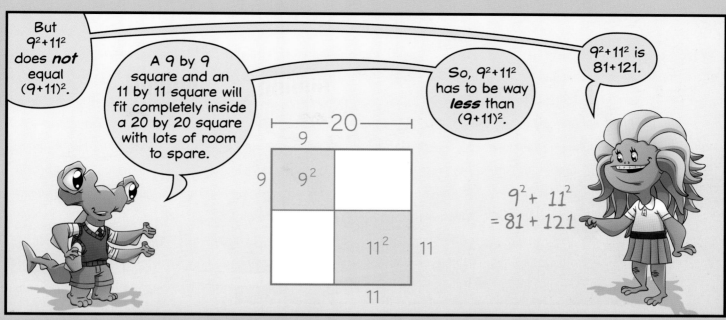

But 9^2+11^2 does **not** equal $(9+11)^2$.

A 9 by 9 square and an 11 by 11 square will fit completely inside a 20 by 20 square with lots of room to spare.

So, 9^2+11^2 has to be way **less** than $(9+11)^2$.

9^2+11^2 is $81+121$.

$9^2+11^2 = 81+121$

BZZZZT!

202.

Correct! Bots lead, 2 to 1.

Question 3: What is the units digit of 17^{16}?

Try it.

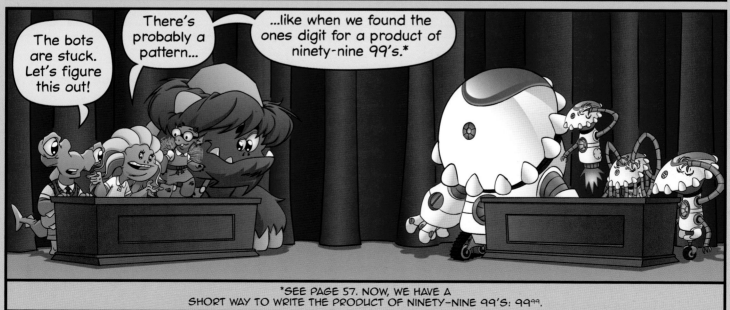

The bots are stuck. Let's figure this out!

There's probably a pattern...

...like when we found the ones digit for a product of ninety-nine 99's.*

*SEE PAGE 57. NOW, WE HAVE A
SHORT WAY TO WRITE THE PRODUCT OF NINETY-NINE 99'S: 99^{99}.

17×17=289, which ends in a 9.

But we only need to multiply the ones digits. 7×7=49 ends in a 9, so 17×17 ends in a 9.

Good. I'll make a chart. 17 ends in a 7, and 17^2 ends in a 9. What about 17^3?

When we multiply a number that ends in 9 by a number that ends in 7...

...we get a number that ends in 3.

Because 9×7=63.

So, 17^3 ends in a 3.

Next is 17^4.

	units digit
17^1	7
17^2	9
17^3	3
17^4	
17^5	
17^6	

17^3 ends in a 3, so if we multiply 17^3 by another 17, we get a number that ends in 1.

Because 3×7=21.

	units digit
17^1	7
17^2	9
17^3	3
17^4	1
17^5	
17^6	

And 17^5 ends in a 7, since 1×7=7.

	units digit
17^1	7
17^2	9
17^3	3
17^4	1
17^5	7
17^6	

Now we're back to 7, right where we started!

What is the units digit of 17^6? 17^7? 17^8?

88

We already know that multiplying a number that ends in 7 by 17 gives us a number that ends in 9.

And multiplying a number that ends in 9 by 17 gives us a number that ends in 3.

Then, multiplying a number that ends in 3 by 17 gives us a number that ends in 1.

	units digit
17^1	7
17^2	9
17^3	3
17^4	1
17^5	7
17^6	9
17^7	
17^8	
17^9	

	units digit
17^1	7
17^2	9
17^3	3
17^4	1
17^5	7
17^6	9
17^7	3
17^8	
17^9	

	units digit
17^1	7
17^2	9
17^3	3
17^4	1
17^5	7
17^6	9
17^7	3
17^8	1
17^9	

The units digit repeat!

7, 9, 3, 1, 7, 9, 3, 1...

	units digit
17^1	7
17^2	9
17^3	3
17^4	1
17^5	7
17^6	9
17^7	3
17^8	1
17^9	7
17^{10}	9
17^{11}	3
17^{12}	1
17^{13}	7
17^{14}	9
17^{15}	3
17^{16}	1

The units digits of the powers of 17 repeat in groups of four. 17^4, 17^8, 17^{12}, and 17^{16} all end in...

1!

Correct! The Little Monsters tie the score at 2.

Question 4: What power of 6 is equal to $6^4 + 6^4 + 6^4 + 6^4 + 6^4 + 6^4$?

Try it.

THE ACTUAL VALUE OF 17^{16} IS 48,661,191,875,666,868,481! TO GIVE YOU AN IDEA OF HOW BIG THAT NUMBER IS, THE DISTANCE FROM THE SUN TO THE NEAREST NEIGHBORING STAR IS ABOUT 48,661,191,875,666,868,481 MILLIMETERS. (THE SIDE LENGTH OF THIS TINY SQUARE ➡▪ IS 1 MILLIMETER.)

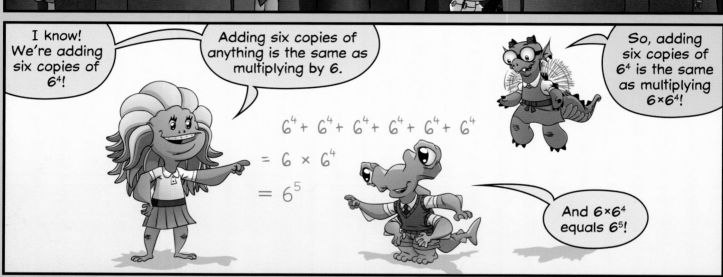

25⁵ means we multiply five 25's.

We need to know how many 5's must be multiplied to equal the product of five 25's.

$$25^5 = 25 \times 25 \times 25 \times 25 \times 25$$

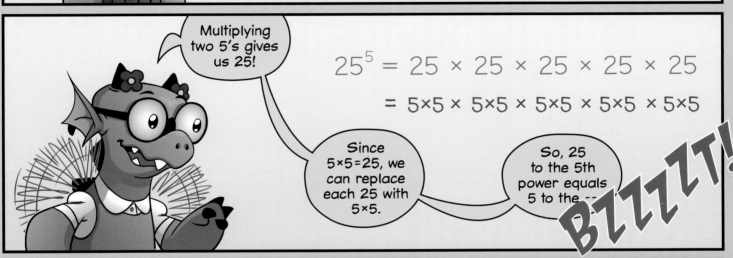

Multiplying two 5's gives us 25!

$$25^5 = 25 \times 25 \times 25 \times 25 \times 25$$
$$= 5{\times}5 \times 5{\times}5 \times 5{\times}5 \times 5{\times}5 \times 5{\times}5$$

Since 5×5=25, we can replace each 25 with 5×5.

So, 25 to the 5th power equals 5 to the --

BZZZZT!

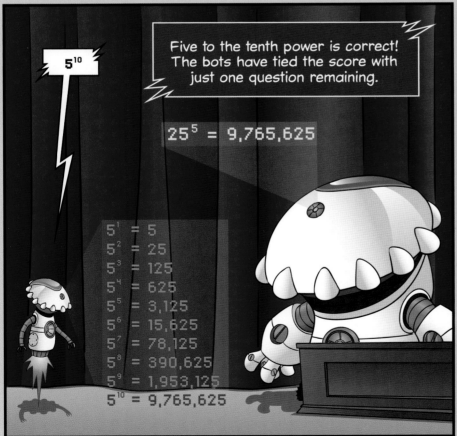

5¹⁰

Five to the tenth power is correct! The bots have tied the score with just one question remaining.

$$25^5 = 9{,}765{,}625$$

$5^1 = 5$
$5^2 = 25$
$5^3 = 125$
$5^4 = 625$
$5^5 = 3{,}125$
$5^6 = 15{,}625$
$5^7 = 78{,}125$
$5^8 = 390{,}625$
$5^9 = 1{,}953{,}125$
$5^{10} = 9{,}765{,}625$

The last question will determine today's math meet winner.

Question 6:
If $c^2 = 121^3$, what is c?

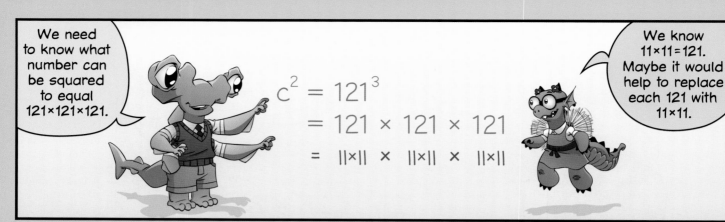

We need to know what number can be squared to equal 121×121×121.

We know 11×11=121. Maybe it would help to replace each 121 with 11×11.

$$c^2 = 121^3$$
$$= 121 \times 121 \times 121$$
$$= 11{\times}11 \times 11{\times}11 \times 11{\times}11$$

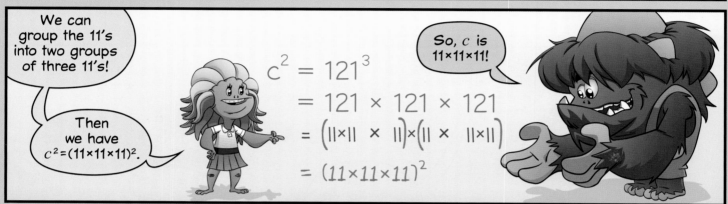

We can group the 11's into two groups of three 11's!

Then we have $c^2 = (11 \times 11 \times 11)^2$.

So, c is 11×11×11!

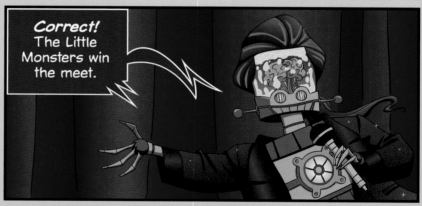

$$c^2 = 121^3$$
$$= 121 \times 121 \times 121$$
$$= \big(11{\times}11 \times 11\big) \times \big(11 \times 11{\times}11\big)$$
$$= (11 \times 11 \times 11)^2$$

11×11=121, and 121×11 equals...

Ding!

...1,331!

$$
\begin{array}{r}
121 \\
\times\ \ 11 \\
\hline
121 \\
+\ 1210 \\
\hline
1331
\end{array}
$$

Correct! The Little Monsters win the meet.

We're goin' to the Math Bowl!

Practice: Pages 81-91

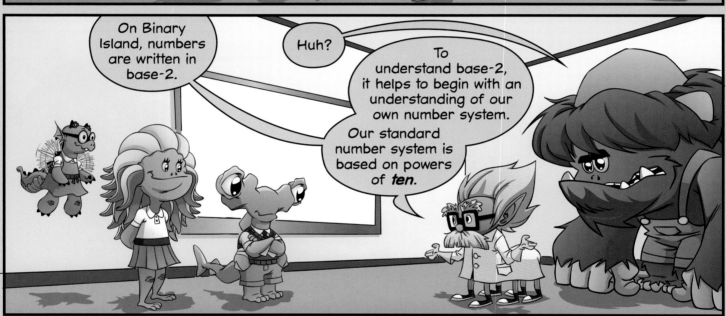

We call our system of numbers **base-10**, because 10 is the base for each place value.

In base-10, we use ten digits, so the place values are powers of 10.

MILLIONS	10^6	=	1,000,000
HUNDRED-THOUSANDS	10^5	=	100,000
TEN-THOUSANDS	10^4	=	10,000
THOUSANDS	10^3	=	1,000
HUNDREDS	10^2	=	100
TENS	10^1	=	10
ONES	10^0	=	1

The powers of ten give us our standard place values: ones, tens, hundreds, thousands, and so on.

$$\overline{10^6}, \overline{10^5}\; \overline{10^4}\; \overline{10^3}, \overline{10^2}\; \overline{10^1}\; \overline{10^0}$$

MILLIONS, HUNDRED-THOUSANDS, TEN-THOUSANDS, THOUSANDS, HUNDREDS, TENS, ONES

WRITING NUMBERS IN BASE-10 IS CALLED *DECIMAL* NOTATION.

On Binary Island, the system of numbers is called **base-2**, because 2 is the base for each place value.

In base-2, only two digits are used, so the place values are powers of 2.

SIXTY-FOURS	2^6	=	64
THIRTY-TWOS	2^5	=	32
SIXTEENS	2^4	=	16
EIGHTS	2^3	=	8
FOURS	2^2	=	4
TWOS	2^1	=	2
ONES	2^0	=	1

So, in base-2, digits don't stand for ones, tens, hundreds, and thousands.

The digits of a number written in base-2 stand for ones, twos, fours, eights, and so on.

$$\overline{2^6}, \overline{2^5}\; \overline{2^4}\; \overline{2^3}, \overline{2^2}\; \overline{2^1}\; \overline{2^0}$$

SIXTY-FOURS, THIRTY-TWOS, SIXTEENS, EIGHTS, FOURS, TWOS, ONES

WRITING NUMBERS IN BASE-2 IS CALLED *BINARY* NOTATION.

So, in base-2, 100 is four, 1,000 is eight, 10,000 is sixteen, 100,000 is thirty-two, and 1,000,000 is sixty-four.

Base-2	Base-10
100	4
1,000	8
10,000	16
100,000	32
1,000,000	64

Very good!

Here is another number written in base-2.

How would we write this number in base-10?

101,110

Try it.

To begin, we can find the place value of each digit.

1 0 1, 1 1 0

THIRTY-TWOS SIXTEENS EIGHTS FOURS TWOS ONES

The base-2 number has 1 thirty-two, 0 sixteens 1 eight, 1 four, 1 two, and 0 ones.

We can add all of these to convert to base-10.

The base-2 number 101,110 equals the base-10 number 46.

1 0 1, 1 1 0

THIRTY-TWOS SIXTEENS EIGHTS FOURS TWOS ONES

$$
\begin{array}{r}
32 \\
8 \\
4 \\
+\ 2 \\
\hline
46
\end{array}
$$

All we need to do is find the number of jellybeans in this jar... in base-2.

Maybe it would help to count how many jellybeans there are in base-10 first.

I can do that!

Don't eat any, Grogg!

...57, 58...

...59, 60, 61...

...62, 63...

Grogg! I already counted that one.

Instead of counting them one-by-one, we should put the jellybeans in piles.

If we make piles of ten, the jellybeans will be easy to count.

There are a lot of piles of ten.

We can combine ten piles of ten to make one pile of one hundred.

That makes one pile of one hundred, one pile of ten, and seven extras.

In base-10, that's 117 jellybeans.

How could they write the number of jellybeans in base-2?

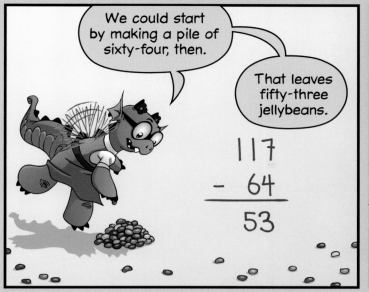

We could start by making a pile of sixty-four, then.

That leaves fifty-three jellybeans.

$$117 - 64 = 53$$

Then, we could make a pile of thirty-two.

That leaves just twenty-one jellybeans.

$$53 - 32 = 21$$

With twenty-one jellybeans, we can make a pile of sixteen and still have five left over.

$$21 - 16 = 5$$

And with the remaining five jellybeans, we can make a pile of four and a pile of one.

sixty-four   thirty-two   sixteen   eight   four   two   one

That gives us 1 sixty-four, 1 thirty-two, 1 sixteen, 0 eights, 1 four, 0 twos, and 1 one.

How do we write the number of jellybeans above in base-2?

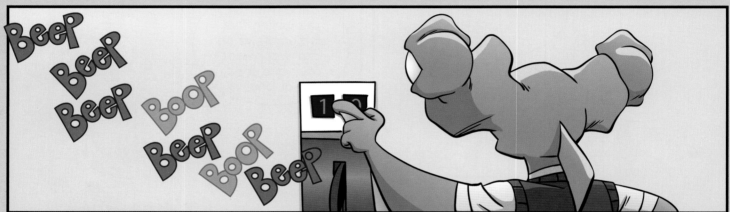

Practice: Pages 92-101

Happy numbers

GrOgg

To find Out if a number is "happy":

1. Square the digits Of the number. Add the squares tO get a new number.

2. Repeat step 1 with the new number until One Of twO things happens:

If yOu get tO 1, the number is a happy number!
If yOu never get tO 1, the number is nOt a happy number.

17
$1^2+7^2=50$
$5^2+0^2=25$
$2^2+5^2=29$
$2^2+9^2=85$
$8^2+5^2=89$
$8^2+9^2=145$
$1^2+4^2+5^2=42$
$4^2+2^2=20$
$2^2+0^2=4$
$4^2=16$
$1^2+6^2=37$
$3^2+7^2=58$
$8^2+5^2=89$
$8^2+9^2=145$
$1^2+4^2+5^2=42$
$4^2+2^2=20$
$2^2+0^2=4$
$4^2=16$
$1^2+6^2=37$
$3^2+7^2=58$
$8^2+5^2=89$
$8^2+9^2=145$
$1^2+4^2+5^2=42$
$4^2+2^2=20$
$2^2+0^2=4$
$4^2=16$
$1^2+6^2=37$
$3^2+7^2=58$
$5^2+8^2=89$
$8^2+9^2=145$
$1^2+4^2+5^2=42$
$4^2+2^2=20$

7
$7^2=49$
$4^2+9^2=97$
$9^2+7^2=130$
$1^2+3^2+0^2=10$
$1^2+0^2=1$

7 is Happy!

1,000
$1^2+0^2+0^2+0^2=1$
1,000 is happy!

(sO is 1,000,000!)

23
$2^2+3^2=13$
$1^2+3^2=10$
$1^2+0^2=1$

23 is happy!

(32 must be happy, tOO!)

$2^2+0^2=4$
$4^2=16$
$1^2+6^2=37$
$3^2+7^2=58$
$8^2+5^2=89$
$8^2+9^2=145$
$1^2+4^2+5^2=42$
$4^2+2^2=20$
$2^2+0^2=4$
$4^2=16$
$1^2+6^2=37$
$3^2+7^2=58$
$5^2+8^2=89$
$8^2+9^2=145$
$1^2+4^2+5^2=42$
$4^2+2^2=20$
$2^2+0^2=4$

17 is nOt happy

Index

For additional books,
printables, and more, visit
www.BeastAcademy.com